ANGRY earth

SAVAGE

TSUNAMIS

By Michael Portman

 Gareth Stevens
Publishing

Please visit our website, www.garethstevens.com. For a free color catalog of all our high-quality books, call toll free 1-800-542-2595 or fax 1-877-542-2596.

Library of Congress Cataloging-in-Publication Data

Portman, Michael, 1976-
Savage tsunamis / Michael Portman.
    p. cm. — (Angry earth)
Includes index.
ISBN 978-1-4339-6551-7 (pbk.)
ISBN 978-1-4339-6552-4 (6-pack)
ISBN 978-1-4339-6549-4 (library binding)
1. Tsunamis. I. Title.
GC221.2.P69 2012
551.46'37—dc23

                                    2011037280

First Edition

Published in 2012 by
Gareth Stevens Publishing
111 East 14th Street, Suite 349
New York, NY 10003

Copyright © 2012 Gareth Stevens Publishing

Designer: Katelyn E. Reynolds
Editor: Therese Shea

Photo credits: Cover, pp. 1, 5 (main), 18, 25 (inset), (cover, pp. 1, 3–32 background and newspaper graphics) Shutterstock.com; (pp. 3–32 text/image box graphics) iStockphoto.com; p. 5 (inset) Wikimedia Commons; pp. 6, 19 iStockphoto/Thinkstock; pp. 7, 21 (inset) AFP/AFP/Getty Images; p. 8 Hiroshi Kawahara/AFP/Getty Images; p. 9 The Asahi Shimbun via Getty Images; p. 10 Sadatsugu Tomizawa/AFP/Getty Images; p. 11 Jeremy Mayes/iStock Vectors/Getty Images; p. 12 Dorling Kindersley/Getty Images; p. 13 DigitalGlobe via Getty Images; p. 14 Colin Rose/Getty Images; p. 15 Mary Plage/Oxford Scientific/Getty Images; p. 16 The Agency Collection/Dorling Kindersley/Getty Images; p. 17 (inset) Jimin Lai/AFP/Getty Images; p. 17 (main) STR/AFP/Getty Images; p. 21 (main) John Russell/AFP/Getty Images; p. 22 (inset) Sena Vidanagama/AFP/ Getty Images; pp. 22 (main), 23 Kazuhiro Nogi/AFP/Getty Images; p. 25 (main) Jiji Press/AFP/Getty Images; pp. 26–27 Toshifumi Kitamura/AFP/Getty Images; p. 29 (inset) Yasuyoshi Chiba/AFP/Getty Images; p. 29 (main) Marco Garcia/Getty Images.

Printed in the United States of America

CPSIA compliance information: Batch #CW12GS: For further information contact Gareth Stevens, New York, New York at 1-800-542-2595.

CONTENTS

Words in the glossary appear in **bold** type the first time they are used in the text.

DESTRUCTIVE HARBOR WAVES

A tsunami is one of the deadliest and most **destructive** forces of nature. The term "tsunami" comes from the Japanese words *tsu* (harbor) and *nami* (wave). A tsunami is actually a series of powerful waves that travel great distances across the world's oceans very quickly. A tsunami wave can be 60 miles (97 km) long and 100 feet (30 m) tall.

The incredible power of a tsunami can destroy almost anything in its path. Cars and trucks, buildings, and even stone structures are no match for a tsunami. These powerful waves have been responsible for countless deaths and billions of dollars in property **damage**.

The Ring of Fire

Most tsunamis occur in the Pacific Ocean. Tsunamis often originate in the "Ring of Fire," an area where underwater volcanoes and earthquakes are common. Tsunamis have also formed in other bodies of water, including the Caribbean and the Mediterranean Seas and the Indian and the Atlantic Oceans.

RING OF FIRE

The series of waves in a tsunami is sometimes called a wave train.

5

TYPES OF WAVES

Most waves are created by wind blowing across the surface of the water. The size and strength of these waves depend on the speed of the wind. Tsunamis, however, are created when a sudden disturbance—such as an earthquake, volcano, or **landslide**—moves a large amount of water.

The size and strength of a tsunami's waves depend on the power of the event that created them. Strong earthquakes and large landslides release great energy that can move, or displace, a lot of water. While the force of the energy pushes the water up, gravity pulls it down. This causes waves to spread in all directions.

Tidal Waves?

Sometimes, tsunamis are mistakenly called tidal waves. Real tidal waves aren't as violent or large. In fact, a tidal wave is another name for the tide. This twice-daily occurrence is caused by the gravity of the sun and the moon pulling on a body of water, causing it to rise and fall.

▲
Swimmers in southern Thailand run to escape
tsunami waves on December 26, 2004.

JET SPEED

While wind-driven waves move in the same direction as the wind, tsunami waves move much like the ripples created by tossing a pebble into a pond. Tsunamis can travel through an ocean at speeds of more than 500 miles (805 km) per hour. This is the same speed that a jet travels through the air. Some tsunamis are even faster. At such high speeds, a tsunami can cross an ocean in just a few hours.

Tsunamis don't lose much energy as they travel through deep water. The energy moves through the water rather than on top as with wind-driven waves. The deeper the water, the faster the tsunami moves.

tsunami hits Sendai, Japan (2011)

Wind Power

Waves created by wind can be very powerful, but they aren't nearly as long, powerful, or fast as tsunami waves. Waves created by the wind usually travel at speeds between 5 and 60 miles (8 and 97 km) per hour.

An earthquake in Chile brought these tsunami waves to the shores of Japan in February 2010.

Wavelength

Tsunamis have long wavelengths. A wavelength is the distance from the top, or crest, of one wave to the crest of the next wave. A tsunami's wavelength may be from 60 miles (97 km) to several hundred miles. It may take more than an hour after the first crest reaches land for the second crest to arrive.

▲

Though not all tsunamis have towering waves, some do. This one hit Minamisoma, Japan, in 2011, killing several hundred people.

In the deep ocean, tsunami waves may rise less than 3 feet (0.9 m) above the ocean's surface. This makes it difficult to spot tsunamis from the air or even in a boat on the water. It's usually only when the tsunami reaches the shoreline that it can be identified easily.

As a tsunami races toward shallow coastal waters, it begins to slow down. A tsunami that was once traveling at 500 miles (805 km) per hour through the deep ocean may slow to 30 miles (48 km) per hour. This sudden change in speed causes tsunami waters to "pile up" and reach heights of up to 100 feet (30 m) above sea level.

▲
diagram showing tsunami
waves increasing in height
near shore

13

EARTHQUAKES

Tsunamis are often the result of large underwater earthquakes. Earth's outer layer, or crust, is composed of huge sections called plates on which the continents and ocean floor rest. These plates move very slowly on a layer of hot, soft rock. When plates meet, pressure builds up between them until they suddenly snap out of place. They may slide past each other, crumble, or slide over or under each other.

The energy released by this sharp movement causes the ground to shake and the water around it to move. Fortunately, most earthquakes don't generate enough force to create a tsunami.

diagram demonstrating how earthquake energy causes a tsunami

This image was taken of Meulaboh, Indonesia, shortly after a tsunami hit in 2004. People fled the shore for higher ground after an earthquake struck, but the tsunami killed many.

▼

Measuring Earthquakes

The Richter scale measures the **magnitude** of earthquakes. The scale ranges from 1 to 10, with 1 as the weakest quake. Generally, tsunamis are caused by earthquakes that are magnitude 7 or higher.

OTHER CAUSES

Volcanoes, landslides, and meteorites are three more causes of tsunamis. When one of Earth's plates slides under another, the sinking plate carries water that lowers the melting point of rock and allows **magma** to form. Two plates pulling apart can also allow magma to rise to Earth's surface. When magma escapes through a volcanic eruption, the explosion can cause a tsunami.

Underwater landslides can cause tsunamis, too. However, since earthquakes trigger most underwater landslides, earthquakes can be considered the true cause.

There are no known meteorite-caused tsunamis in history. However, scientists think a large meteorite crashed into Earth over 3 billion years ago and created a powerful tsunami that swept over the entire planet—several times!

illustration of volcanic activity under the seafloor

Krakatoa

On August 26, 1883, one of the largest and most destructive tsunamis ever recorded occurred. The Indonesian volcano Krakatoa, or Krakatau, exploded, creating tsunami waves that reached 130 feet (40 m) high. Many towns and villages on nearby islands were destroyed, killing about 36,000 people.

◄ Anak Krakatau ("son of Krakatoa") rose from almost the same spot as the killer volcano and continues to grow.

REACHING LAND

When a tsunami reaches land, it causes the most damage to areas within 1 mile (1.6 km) of the shoreline. Sometimes, a tsunami crashes onto shore with a tall, violent wave called a bore. Most often, it looks as if the ocean tide is rising up quickly, or surging, onto the shore.

Without tall waves, the water surface may appear calm, but underneath it's a different story. Powerful, churning water can easily pull people under, cause cars to float like toys, and rip buildings from the ground. Entire beaches can be washed away by tsunamis.

Wave of Noise

Just as waves make a crashing sound, a tsunami's incredible power is extremely loud. Some people have compared the sound to trains or jet engines. When people hear the noise, they should **evacuate** to high ground as quickly as possible.

southwestern coast of Sri Lanka
after 2004 tsunami

▲
This beach is littered with objects that
were carried by tsunami waves.

VANISHING OCEAN

A person on the beach may notice that water on the shoreline moves back and forth. Just before a wave hits the shore, the water drains back into the ocean or lake. The same thing often happens just before a tsunami reaches shore, only on a much bigger scale.

Many times, water along the coastline will completely drain away when a tsunami approaches. This is because the low point of the wave, called the trough, often reaches shore first. When it does, it causes water along the coast to pull back towards the sea, exposing large amounts of seafloor. This vacuum effect warns that the wave's crest will soon follow.

evacuation signs lead people to safety

A drained shore before a tsunami looks much like this area surrounding a dock.

▼

CAUTION

Every Minute Counts

It takes about 5 minutes for the first crest of a tsunami to reach shore after the trough drains the coast. This should alert people along the coastline. Other waves can follow anywhere from 5 to 90 minutes after the first wave hits.

THE 2004 INDIAN OCEAN TSUNAMI

On December 26, 2004, the most powerful earthquake in 40 years shook the seafloor beneath the Indian Ocean. Measuring 9.1 on the Richter scale, this massive quake sent a series of killer waves speeding in all directions across the Indian Ocean. In just a few hours, tsunami waves crashed onto the shores of 11 countries around the Indian Ocean, killing more than 200,000 people.

The tsunami was so powerful that its destructive waves traveled more than 3,000 miles (4,827 km), causing damage along the east coast of Africa. In some areas, the earthquake created tsunami waves that reached 50 feet (15 m) tall.

Explosive Destruction

The Indian Ocean earthquake was caused by two plates that had pressed against each other for hundreds of years. When they suddenly snapped out of place, they released energy that scientists **estimate** was equal to 23,000 **atomic bombs**.

tsunami in Malaysia (2004)

In many places, the Indian Ocean tsunami waves were small. In others, they towered over buildings.

Sri Lanka (December 2004)

This photo shows an area of Indonesia several weeks after the 2004 tsunami. Flooding damaged many buildings beyond repair.

Unfortunately, many people affected by the Indian Ocean tsunami didn't recognize one of the warning signs. Just before it reached land, several coastlines were drained of water, **stranding** boats and fish. Because this was such a rare sight, many curious people wandered onto the bare seafloor. A few minutes later, the tsunami waves rushed onto the shoreline, swallowing everything in their path.

The Indian Ocean tsunami destroyed thousands of miles of coastline and flooded entire islands. By the time the water level returned to normal, hundreds of thousands of people were dead or missing, and millions more were homeless.

Lifesaving Knowledge

A man in India saved his village when he remembered what he learned from a TV show about tsunamis. A young girl saved her family recalling what she learned in science class. Many animals sensed the danger, too. People reported seeing them run to higher ground just before the tsunami arrived.

THE 2011 JAPAN TSUNAMI

On March 11, 2011, a massive earthquake struck off the coast of Japan's main island, Honshu. This quake measured 9.0 on the Richter scale. The tsunami it created **devastated** parts of Japan and also caused destruction in other countries along the Pacific Ocean. Giant waves flooded cities and villages, swept away cars and trucks, and destroyed homes and businesses. Even the West Coast of the United States suffered damage from the tsunami.

When tsunami waters flooded a Japanese power plant, deadly **radioactivity** was released into the water and air. No one knows when the area will be safe to live in again.

A History of Earthquakes

Japan has suffered many earthquakes throughout its history. The 2011 earthquake, however, was the largest to ever strike the country. Yet the earthquake caused only a small fraction of the damage, thanks to earthquake-ready buildings. Most of the damage was caused by the tsunami.

location of 2011 earthquake near Japan

▲
This picture was taken by an official in Miyako, Japan, on March 11, 2011. It shows the tsunami flowing into the city. Notice the wave's crest isn't towering but rather creates a powerful surge onto the shore.

25

UNPREDICTABLE WAVES

There's currently no way to **predict** when a tsunami will occur. Earthquakes, landslides, and volcanic eruptions also cannot be predicted and don't always result in a tsunami. In addition, once a tsunami begins, it's only a matter of minutes or hours before it reaches land.

However, scientists have placed **sensors** on the ocean floor to detect activity. Once movement occurs, they use computers to create models that predict wave size and the time of their arrival on land. Scientists also use sea-level **gauges** to help identify tsunamis. Tsunami warning centers use all this information to provide warnings to the public.

Some countries, such as Japan, have built concrete walls to keep tsunami waves from reaching the shore. Unfortunately, those walls weren't strong enough or tall enough to protect many areas from the damage of the 2011 tsunami.

Future Quake?

The coast of the northwestern United States and southwestern Canada is located near two plates capable of creating powerful earthquakes. It has been roughly 300 years since an earthquake similar to the one that struck Japan in 2011 has occurred there. An earthquake of that size could create a deadly tsunami.

TSUNAMI SAFETY

The best way to protect yourself against a tsunami is to act quickly after receiving a warning. If you live in an area that may be affected by a tsunami, be sure to get to higher ground as soon as you hear a tsunami warning, feel an earthquake, or see the coastal waters draw back.

It's also important to remember that a tsunami is a series of waves. The waves that follow the first can be even larger, and surging waters are equally dangerous. When it comes to tsunamis, every minute counts.

More Tsunami Facts

- Tsunamis can travel up rivers and streams that lead to the ocean.
- The Pacific Tsunami Warning System is stationed in Hawaii.
- Eighty percent of tsunamis begin in the "Ring of Fire."
- Tsunami waves can bounce repeatedly off the edges of bays and harbors, causing the waves to grow larger.
- Because tsunami waves can be tall, they may affect inland areas that are less than 50 feet (15 m) above sea level.

Hidden Dangers

Floodwaters caused by tsunamis can contain harmful chemicals and waste. Avoid eating food or drinking water that has come into contact with floodwater. People should follow the orders of officials before, during, and after a tsunami.

A scientist at the National Weather Service Pacific Tsunami Warning Center monitors computer-tracking systems watching for tsunami activity in the Pacific Ocean.

GLOSSARY

atomic bomb: a bomb whose explosive power comes from energy released by splitting atoms

damage: harm. Also, to cause harm.

destructive: causing ruin

devastate: to cause widespread damage

estimate: a guess based on collected facts

evacuate: to withdraw from a place for protection

gauge: a tool that measures the amount of something

landslide: the sudden movement of rocks and dirt down a hill or mountain

magma: liquid rock deep within Earth

magnitude: a measure of the power of an earthquake

predict: to guess what will happen in the future based on facts or knowledge

radioactivity: a harmful kind of energy

sensor: a tool that can detect changes in its surroundings

strand: to cause to remain behind

FOR MORE INFORMATION

Books

Aronin, Miriam. *Slammed by a Tsunami!* New York, NY: Bearport Publishing, 2010.

Roza, Greg. *The Indian Ocean Tsunami.* New York, NY: PowerKids Press, 2007.

Walker, Niki. *Tsunami Alert!* New York, NY: Crabtree Publishing, 2006.

Websites

FEMA: Tsunami
www.fema.gov/hazard/tsunami/
Read what to do before, during, and after a tsunami on the Federal Emergency Management Agency's website.

How Tsunamis Work
science.howstuffworks.com/nature/natural-disasters/tsunami.htm
Learn more about how tsunamis form and about the worst tsunamis in history.

Publisher's note to educators and parents: Our editors have carefully reviewed these websites to ensure that they are suitable for students. Many websites change frequently, however, and we cannot guarantee that a site's future contents will continue to meet our high standards of quality and educational value. Be advised that students should be closely supervised whenever they access the Internet.

INDEX